R. J. WILKINSO

Discovering
Artillery

SHIRE PUBLICATIONS LTD

CONTENTS

The cover illustration is part of 'Peace Manoeuvres: A Field Battery, Royal Artillery, coming into action, supported by Infantry of the Line' by R. Simpkin.

ACKNOWLEDGEMENTS
I would like to thank the following who have helped in the preparation of this book: the late Major R. G. Bartelot of the Royal Artillery Institution, Woolwich, and the staff of the Rotunda Museum; Jack Cassin Scott, who besides undertaking research work did a number of the line drawings for the text and F. Hinchcliffe for information on model making. Illustrations are acknowledged to: the Admiralty, plate 5; Jack Blake, plates 4, 13 and 29; Cadbury Lamb, plates 6, 9, 17, 19, 20, 21, 25, 26 and 28; Royal Artillery Institution, plates 2, 4 and 29.

Printed in Great Britain by C. I. Thomas & Sons (Haverfordwest) Ltd, Press Buildings, Merlins Bridge, Haverfordwest, Dyfed.

1. INTRODUCTION: A GENERAL HISTORY TO 1650

The early history of cannon and gunpowder is extremely vague, but we do know that cannon were in existence by 1326 as they are mentioned in a decree of the city of Florence for that year. Also, a manuscript of the same date, in the British Museum, entitled *De Notabilitantibus Sapiens et Prudentia Regum,* by Walter de Milimete, shows one of these early weapons being fired by a soldier in chain-mail armour. The weapon being fired is vase-shaped and was called a *vasi* by the Italians and a *pot-de-fer* by the French. It was cast in bronze and the projectile fired was an arrow, which had the shaft tightly bound with leather so that it fitted tightly into the bore. The propellant, which was gunpowder, was ignited through a small touch hole in the breech end by a red-hot wire or iron.

All the known evidence concerning the discovery of gunpowder points to the friar Roger Bacon (1214-1292) as the first man to record the formula and properties of gunpowder, but evidence is lacking to attribute the discovery of the propellant itself to Bacon. Many names have been put forward over the years by a number of authors and historians as the possible discoverer of gunpowder but conclusive proof is lacking in every case. What we do know, however, is that gunpowder and cannon were in use in Europe early in the fourteenth century.

Because of the crude methods of casting used at the time, the *pot-de-fer* did not increase much in size. Instead, the larger pieces were made from iron bars and hoops, hammer-welded together. There are a number of examples of this type of cannon still in existence in museums in England and elsewhere.

Edward III used 'crakys of war' against the Scots in 1327 but these were probably firecrackers which frightened by their noise rather than cannon which fired projectiles. The new form of artillery, the cannon, did not immediately become widespread because of the individual cost of each piece and each round fired as well as the uncertainty of using it. A more important reason was that the power of these new weapons was not yet sufficient to replace the trebuchet, a machine like a Roman *ballista* which hurled large stones at fortified positions. The English used cannon at the battle of Crecy in 1346 but we do not know how effective they were. It is more than probable that any success they had was due more to the noise than to the projectile fired.

The French had used artillery as early as 1338 and the Italians had had them before this date as we have already seen. The bombards (from the Greek *bombos,* meaning a noise) that were used at Crecy were made from hoops of iron hammer-welded on to iron bars. The finished barrel, which tapered towards the breech end, was mounted on a crude wooden carriage by means of large iron bands which fitted at each side. The carriages were known as 'great trunks' because they were little better than trunks of trees. The first real effective use of cannon was by the English at the siege of Calais in 1347.

At this stage in the development of guns, there comes a diversification. The smaller cast bronze *pot-de-fer* had by now been mounted on a wooden shaft which could be operated by one man, with the end of the shaft dug into the ground and a rest for the barrel, whereas the larger iron guns became even bigger and heavier as the walls of towns and castles thickened to combat this new form of warfare. In this book we leave the 'handgonne' as it became known and concentrate on the further development of the larger types of artillery.

The siege gun was very cumbersome and heavy, and the carriages matched their barrels for weight and awkwardness. The basic carriage consisted of a heavy piece of timber with the barrel firmly strapped to it with iron bands (fig. 1), and it was elevated, when required, by putting blocks of wood under the barrel end until the desired angle was reached. For the lighter pieces, which were a little heavier than the 'handgonne', various types of carriage existed, but none to a uniform design. Most employed a wooden platform, sometimes with small, round, wooden wheels, called trucks, on which was mounted the elevating gear (fig. 2), in which the barrel fitted. A number of these early guns were breech-loaders; the chamber holding the projectile and charge was removable, so that a number of pre-loaded chambers could be kept ready, thus speeding the rate of fire (fig. 3). The chamber was slotted into position behind the barrel and held there tightly by having a wooden wedge driven in behind the chamber and between it and the back stop of the carriage (fig. 4).

Some of the carriages had wheels but for the most part cannon were carried on wagons. During sieges of towns and other fortified places, many of the pieces were equipped with shields to protect the gunners from enemy arrows, while the

4

 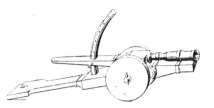

1. *A cannon mounted on its carriage with iron bands.*
2. *Italian bombard of c. 1500 showing the early form of an elevating arc.*
3. *An early breech-loader showing the chambers.*

gunners of smaller pieces without shields constructed rudimentary defences once the gun had been positioned.

The first projectiles fired from these cannon were arrows but these were soon replaced by stone balls, although as late as the early 1400s, arrows were still in store at the Tower of London for use in cannon! The main problem with stone balls was that they were found to break up on impact and as early as 1350 cast iron shot was tried while in the 1370s lead shot, or 'pelottes', was being made at the Tower. The great expense, the difficulties in casting, and the increase in the powder charge necessary to achieve the same range as with stone shot delayed the introduction of iron shot until the middle of the fifteenth century.

Solid stone or cast iron balls were all very well against solid objects such as the walls of towns and castles but there was no effective projectile that could be used against troops in the open. The invention of case shot, which was a can full of small iron or lead pellets, and langridge shot, a similar can filled with scrap iron and stones, provided an answer. The origin of the name langridge is unknown but we do know that this form of projectile was used at the siege of Constantinople in 1453.

The emphasis was now on bigger and bigger guns which could hurl larger projectiles of stone or cast iron against the thick walls of fortified positions, thus giving rise to some

wedge ➤

4. *A built-up gun showing the chamber, wedge and carriage of the type used at sea.*

monster weapons which still survive today, such as 'Mons Meg' (now in Edinburgh Castle), so called because it was said to have been made at Mons, the Dardanelles gun (now in the Tower of London), named after the place it was sited at and capable of firing across the straits, and 'Queen Elizabeth's Pocket Pistol', a large cast bronze gun (now in Dover Castle), that was actually given to Henry VIII. Both 'Mons Meg' and the Dardanelles gun are breech loaders in which the rear part of the barrel was unscrewed, prior to loading, by levers placed in slots. The former is a gun made up of bars and hoops, while the latter is cast.

The development of artillery was further accelerated by the introduction of trunnions, round projections from each side of the barrel at the point of balance, which enabled the barrel to be elevated, mounted, and unmounted more quickly.

As artillery became a science it was soon realised that the use of larger and larger guns was not the real answer. The difficulties of transportation, together with the results achieved, soon made it apparent that the performance did not justify the enormous cost of these pieces. Advances in casting enabled the entire barrel to be made in one rather than assembled from bars and hoops hammer-welded together, and so allowed faster and cheaper construction of guns, making it possible for an army to have more fire power in the field. The first cast iron guns made in England were cast by Ralphe Hogg at his foundry in Sussex in 1542. These guns were cast complete with the bore which had to be cleaned out before being ready for use (see Chapter 5).

Another early form of artillery was the mortar, which had a very short barrel but fired large projectiles at high angles. They were first used in about 1400 and by 1460 had acquired trunnions. The high angle of fire enabled the gunners to lob solid shot or incendiary shells over the walls of besieged towns in the hope of destroying or setting fire to the buildings inside. There were two ways of firing explosive or incendiary shells from a mortar. The shell, containing either a charge of explosive powder or some combustible material, could be loaded with the fuse towards the powder charge, so that when the main charge was ignited the blast lit the fuse on the shell. Alternatively the main charge could be lit slightly after the fuse, placed facing outwards on the shell, was ignited.

Henry VIII established the first permanent force of artillery in England when he appointed a Master Gunner and twelve gunners, to be stationed at the Tower of London. Despite this advance the train of artillery still had to be formed in time of war and men had to be recruited to work the guns, but at least the Master Gunner and his twelve assistants were a nucleus around which to form the train. The train consisted of the pieces of ordnance, guns and mortars, together with a large assortment of wagons and carriages. The horses or other draught animals and their drivers were provided by a civilian contractor. The disadvantages of this system are obvious but it nevertheless continued until 1716 when the first permanent artillery was formed.

In this period guns were popularly named after saints or other famous people, and Henry VIII and Charles V of Spain each had a set of twelve guns named after the twelve apostles. Henry's 'St. John' became stuck in the mud at the siege of Terouenne in 1513 and was captured by the enemy! The various types of cannon, whether guns or mortars, were mostly named after birds of prey or other creatures with nothing in their name to indicate their calibre or performance. The train at the siege of Boulogne, for instance, consisted of 10 cannon, 11 demi-cannon, 21 culverins (fig. 5), 14 demi-culverins, 13

5. *A grand culverin complete with rammers etc. of 1620.*

falcons, 20 sakers, 5 bombards, 1 cannon-perrier, 50 mortars, 50 shrimps and 17 small falcons. The list at the end of the chapter gives details about the various types of cannon. The only item mentioned in the list of the siege train that was not really a cannon was the shrimp, which was more of an armoured car, having small-calibre guns mounted on a two-wheel carriage protected by a shield. It was pushed towards the enemy and fired when the range was right.

The manufacture of the barrel and its carriage was a lengthy procedure and involved making models from which moulds were taken. This is fully dealt with in Chapter 5.

The absence of roads and the heavy weight even of the field guns hindered the tactical use of artillery in this period. So bad were the conditions at times that in some cases the artillery either failed to arrive before the battle was over or, arriving in time, took so long getting into position and ready to fire that the guns hardly influenced the outcome of the conflict. Mobility was not the key word with artillery and guns once positioned before a battle never moved, unless there was danger of their capture.

Artillery was not used as a mobile weapon because the men who manned the guns were on foot and had to march with the pieces, though its use as a mobile weapon had been tried with some success by Marlborough at the battle of Blenheim in 1704, and later at the battle of Fontenoy in 1745. At Fontenoy use was made of 'gallopers' or galloper guns, 1½ pdrs mounted on light carriages with split trails, rather like farm carts, which could be pulled by one horse (plate 1). They could keep up with cavalry and provided highly mobile fire power, but in the following year, 1746, for no given reason, they were discontinued.

Light mobile artillery was used on the Continent many years before its introduction to England and it was the King of Prussia who first set the standard for artillery used in mobile warfare which other countries were quick to follow. The British answer to the use of mobile artillery was battalion guns, allotting to each infantry battalion a number of light guns that moved up with it (see Chapter 2).

The state of the artillery at the start of the Civil War in England is shown in *The Gunners Dialogue* by Robert Norton, published in 1643:

	weight in lbs.	calibre in ins.
Cannon of 8	8000	8
Cannon of 7	7000	7
Aspicke	7600	7½
Demi-cannon	6000	6½
Culverin	4500	5½
Pelican	2550	4¾
Demi-culverin	2500	4½
Saker	1500	3½
Minion	1200	3¼
Falcon	700	2¾
Falconet	500	2¼
Cannon-perrier	3500	9–11
Demi-cannon drake	3000	6½
Culverin-drake	2000	5½
Demi-culverin-drake	1500	4½
Saker drake	1200	3½

Besides these Norton notes a syren at 8,100 lbs, a sparrow at 4,600 lbs, a dragon at 1,400 lbs and a base at 450 lbs.

The average ranges of these pieces were given by Norton: for example, the culverin at an elevation of 10 degrees could fire a distance of 2,650 yards, and a falcon at the same elevation could achieve a range of 1,920 yards. It is very difficult to ascertain much information about the performances of the pieces as it was not until the middle of the 1600s that artillery became a science and books of an instructional nature were published.

2. FIELD ARTILLERY

1650-1770

On 15th February 1645 the New Model Army, under the command of Sir Thomas Fairfax, came into being. The infantry and cavalry were completely reorganised but the artillery, which was considered of relatively minor importance, was left unchanged. The train still had to be formed in time of need around the nucleus of gunners and their mates who were stationed at the various coastal forts and towns of England. Drivers and horses still had to be procured from civilian contractors and the gunners still marched on foot to the place of battle.

One of the most important improvements to field guns came about in 1680 when limbers were introduced to England, though they had been used previously, mainly by the Prussians. The limber was a two-wheeled wagon or just an axle with two wheels to which the gun carriage was fixed by the metal eye in the end of the trail (figs. 6, 7 and 8). The advantage of the limber was that it gave the gun increased mobility and allowed the draught animals to be easily harnessed to pull the combination rather than the cumbersome method of harnessing the animals directly to the trail of the gun. Figs. 7 and 8 show the limber system as used by the Prussians in 1636.

6. *Cannon and draught horses showing the use of an early form of limber.*

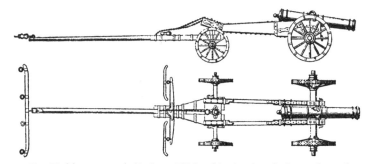

7. *Field gun and limber 1636; this is the design adopted first by the Prussians.*
8. *Elevation of a field gun and limber, 1636; this shows the design used for over 200 years.*

Limbers were soon adopted by most countries although it was to be a further 100 years before ammunition boxes were fitted to limbers.

The basic field carriage of 1680 (fig. 9) did not alter very much in the next 100 years or so. It comprised of wheels, axle

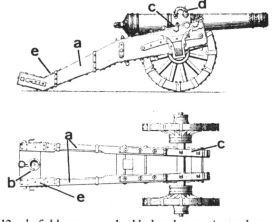

9. *12 pdr field gun on a double bracket carriage whose style changed little over 100 years: a, trail; b, trail eye for attaching to limber; c, capsquares; d, dolphins, or lifting handles on barrel; e, metal reinforcing bands and plates.*

10

and trail. The trail (fig. 9a) was made from two pieces of timber, suitably shaped and held together by the axletree and a number of cross struts along its length. At the end of the trail was the eye (fig. 9b), reinforced with metal, that fitted to the limber. The two stout pieces of timber were known as 'cheeks' and were shaped according to the type and weight of barrel that would be mounted on the carriage. In places where there was likely to be a lot of wear or strain, the wood was bound and reinforced with iron bands and plates were screwed to the carriage (figs. 7 and 8).

The axletree was also made of wood and was fitted to the cheeks by iron bolts which passed through them and were secured to the iron plate at the top. On this iron plate were fitted the capsquares (fig. 9c), which were made of iron and were constructed to hold the barrel firmly in place on its carriage by means of the trunnions (fig. 10). The wheels (fig. 11) were constructed around the hub or nave with the spokes fitted into mortices in the nave, but set at an angle.

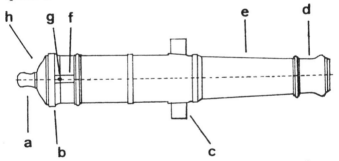

10. *Names of the parts of a barrel: a, button; b, breech; c, trunnions; d, muzzle; e, chase; f, vent field; g, vent; h, cascable.*

This angle was known as the 'dish' of a wheel. Wheels, whether for a gun carriage or an ordinary farm cart, were all dished as this gave the wheel the strength to withstand any inward pressure it might receive when going along. The axletree arms were also inclined at an angle so that the spokes of the lower part of the wheel would be perpendicular to the ground. This was known as the 'hollow'.

To the end of each two spokes was fitted a felly or felloe, so that in a wheel there were six felloes which formed the inner wooden tyre. Once all the felloes were in position the

11

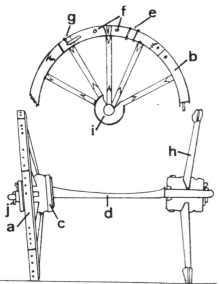

11. *Typical construction of a field carriage wheel: a, tyre streak; b, felloe; c, stock hoop; d, axletree; e, tyre bolts; f, rivets; g, tyre nails; h, spoke; i, nave or hub; j, linchpin.*

iron tyre was heated until red hot and placed on the wheel. When in position it was sprayed with water so that the metal shrank and formed a tight fit with the wooden tyre. In the artillery of the eighteenth and nineteenth centuries the tyres were made up of sections, each fitted to the wheel with bolts and nails. This method had the advantage over the shrunken-on complete tyre that if the wheel was damaged in battle the damaged section only of the tyre need be repaired. (See fig. 11 for the names of the parts of a wheel.)

Although of similar construction, the carriage for the howitzer, a cross between a gun and a mortar capable of firing a heavier projectile than the gun and more mobile than the mortar, was shorter and slightly heavier (fig. 12). The howitzer was first used in British service in about 1720. Each carriage, whether for gun or howitzer, although similar in appearance, differed in weight and measurements according to the type of barrel that was mounted on it.

Even with the new-found mobility of gun limbers, the artillery train still continued to move at the same pace, because the gunners still marched on foot. The civilian drivers supplied by the contractor when the train was formed were often found to be unreliable, many deserting and leaving the guns stranded when the fighting became a bit difficult, so that James II formed the Royal Fusiliers in 1685. The Fusiliers' job was to march with the artillery train and to instil discipline into the drivers.

In an inventory of the arms and stores in the Tower of London for the year 1675, the pieces in store were many and varied, some being referred to by their old names, e.g. falcons and murderers (a small mortar), while others were described as 24 pdr, 12 pdr etc., which was the weight of the largest round shot they could fire. Towards 1700 the names were discarded and ordnance became known by the weight of the shot, except for mortars which were designated in inches.

The methods of loading and firing had hardly improved since early times. The charge of powder was ladled from the barrel, placed in the muzzle of the gun and pushed down to the end of the bore. This was followed by a wad which was rammed up against the powder. The projectile was then placed in the bore and rammed home hard. During this operation the ventsman, the gunner whose job it was to prime the vent with loose powder that was ignited and which in turn fired the main charge, kept his thumb over the vent. This prevented any rush of air caused by the quick movement of the rammer from igniting any stray particles remaining from the last round, which would light the new charge of powder being introduced and so cause an explosion. Once the gun was loaded and the gunners were in position to fire (plate 2), the linstock, a pole as long as a short pike with an attachment to hold the burning match (fig. 13), would be got ready. When the order was given the burning match was placed in the loose powder on the vent. As soon as this ignited it sent the flame

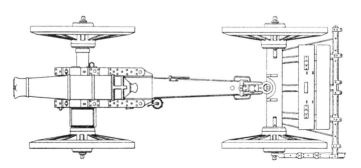

12. *32 pdr howitzer and limber.*

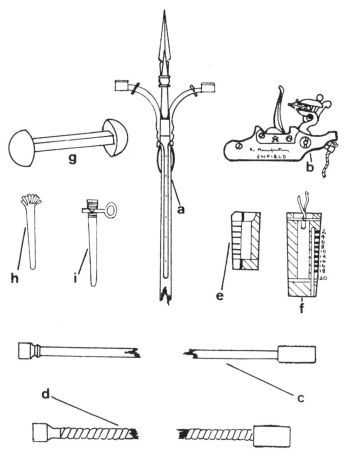

13. *Projectiles, fuses and firing gear: a, linstock; b, gun lock;*
c, wooden shaft rammer; d, rope shaft rammer; e, wood
fuse; f, Boxer's improved fuse; g, bar shot; h, quill tube;
i, friction tube.

down the vent hole to the powder in the bore, and in an instant
the main charge exploded and fired the projectile. By about
1680, the portfire, a short stick which held the burning match,
had replaced the linstock. The linstock was still used, placed

14

in the ground between two guns, as a ready source of fire for the portfire.

By 1700 artillery was becoming an exact science and books were published giving detailed tables of performances of the various guns and mortars in service. As has been said, guns were now known by the weight of round shot they fired and this new method was found to be better than the former one of naming pieces after birds of prey etc.

The main disadvantage of artillery at this time was delay in arriving at an appointed position. This is not surprising when one considers that a siege train then consisted of about 100 guns, 60 mortars and more than 3,000 wagons and 15,000 horses together with the drivers and gunners. The length of a train such as this on the march would be about 15 miles.

Following the peace of Utrecht in 1713, the artillery train was disbanded as it always was once a peace had been signed, and the contractors took back their men, horses and wagons. The formation of a permanent artillery, as opposed to the gunners stationed at the Tower of London and other forts, owes a lot to the 1715 Jacobite rebellion, when the train had to be formed to fight the rebels. The time taken in forming the train was such that the rebellion was over before the train was ready to march. In May 1716 two companies of artillery were formed and stationed at Woolwich. In 1722 these companies were augmented and by 1744, the 8th Company had been formed of officer cadets. By 1757 there were sufficient companies to form two battalions and further battalions were added so that by 1771 there were four battalions of artillery.

May 1716 was an important date for the artillery not only because it was permanently formed then but also because a disastrous explosion at the only brass gun foundry in England at Windmill Hill in the City of London persuaded the Ordnance to set up its own brass foundry, independent of outside contractors, at Woolwich. This became the Royal Brass Foundry. The improved methods of manufacture adopted there are dealt with in Chapter 5.

A most important part of any piece of ordnance was the markings to be found on the barrel. These normally consisted of the crown and cipher of the reigning monarch, although earlier examples bear the entire coat of arms, and the initial and the coronet appropriate to his rank of the Master General of the Ordnance. This crest is most essential to the model maker, as it helps to date the gun. A model of the type of gun used at Fontenoy would hardly be correct with the initial and coronet of John, Marquis of Granby, who was Master General from 1763-70.

By 1750 there was a large variety of both brass and iron guns, howitzers and mortars in service and these altered very little in the next fifty years, although there was much experimenting. The main reason for the disregard of many of these ideas was because the Board of Ordnance was hesitant in replacing pieces in service, which had shown themselves to be reliable and cheap to make, with any untried and probably expensive substitute. This state of affairs was to last until the advent of William Congreve who implemented many new designs in artillery (see 1770-1855).

The brass pieces of ordnance in use with the artillery between 1750 and 1770 are set out below (in each calibre there was a variety of lengths and weights).

Guns: 42 pdr, 32 pdr, 24 pdr, 18 pdr, 12 pdr, 9 pdr, 6 pdr, 3 pdr, and 1 pdr. (Although the 9 pdr was first cast in 1719, it was not used between 1750 and 1808.)
Howitzers: 10 ins., 8 ins., heavy $5\frac{1}{2}$ ins., light $5\frac{1}{2}$ ins., and 4 2/5 ins.
Mortars: 13 ins., 10 ins., 8 ins., $5\frac{1}{2}$ ins., and 4 2/5 ins.

The following iron pieces of ordnance were in use with the artillery between 1750-70:
Guns: 42 pdr, 32 pdr, 24pdr, 18 pdr, 12 pdr, 9 pdr, 6 pdr, 4 pdr, and 3pdr.
Mortars: 13 ins., 10 ins., and 8 ins.

1770-1855

The period 1770-1855 saw the muzzle-loading ordnance rise to the height of perfection in both construction and use. This period is notable for the introduction of many new designs and improvements in existing patterns, and for the removal from service of some of the larger brass guns and howitzers because of their lack of success in the Peninsular War. The main improvements of the period were the block trail field carriage and the carronade (see Chapter 4, 1770-1855).

The block trail field carriage was designed by William Congreve and brought into service in 1792. The advantage of the block trail was that it did away with the two cheeks of the carriage and had instead two shorter cheeks fitted to a single solid wooden trail (fig. 14). This new arrangement made limbering and unlimbering easier and gave greater mobility and speed in turning and manoeuvring. The block trail was first introduced on the 6 and 9 pdr guns and later was adopted for the 12 pdr gun and 12 and 24 pdr howitzers. It was not

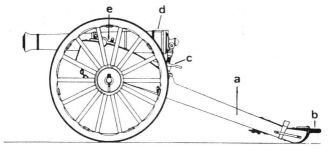

14. *Typical brass field gun on a Congreve block trail, 1796:*
a, trail; b, trail eye; c, elevating gear; d, vent; e, capsquare.

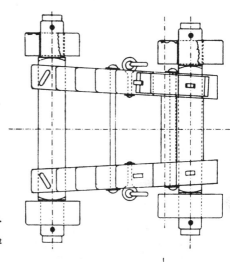

Typical construction
of a garrison carriage.

Jack Cassin Scott

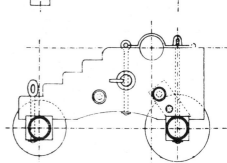

universally adopted until after the Crimean War, the larger calibre guns and howitzers still being mounted on double bracket carriages.

A further improvement made to garrison gun carriages (fig. 15) was the introduction of a new type of quoin, an elevating wedge which was used to raise or lower the barrel. The new quoin now had a horizontal screw running through it which either expanded or contracted the two halves of the quoin and so raised or lowered the barrel. This system was replaced on field carriages by a capstan-headed screw (figs. 14 and 16) which was fitted to the cascable of the barrel (figs. 10 and 14) and passed through the trail or trail bracket.

The normal garrison carriage that was used to mount pieces of various calibre in the forts and coastal defences of Britain and the Empire was different from that used at sea only in the design of its wheels. On land iron wheels were used, whereas at sea these wheels or trucks were of wood (fig. 15 and plate 3). (For a description of garrison carriages, see Chapter 4.) One type of garrison carriage peculiar to land artillery was the iron carriage (plate 4). These were never used in the Navy and on land only in time of peace as the iron carriages tended to break and splinter when hit by an enemy projectile. So for every iron carriage of either guns or carronades used on land there was always a wooden one in store in case of war. Carronades (plate 5) are usually associated with the Navy, but a large number were used on land in forts and coastal defences. The advantages of the carronade were its short barrel, its power at close range and its easy loading.

One unique type of carriage should be mentioned here. This was the one designed by Lieutenant G. F. Koehler of the Royal Artillery during the siege of Gibraltar by the Spanish in 1781 and known as a Gibraltar depression carriage (fig. 17). It was found during the siege that many of the fortress guns could not be depressed sufficiently to hit the floating batteries which lay at sea just off the rock and that the British shots were going over their target. The new invention allowed the gunners to depress the barrel a considerable way below the horizontal so that they could engage the Spanish floating batteries. On 13th September 1782 the guns of the fortress destroyed the floating batteries and inflicted many casualties on the Spanish fleet. When using the depression carriage the powder and shot had to be rammed home well and wadded to prevent the projectile from rolling out before it was fired.

Normal gun carriages used in forts were called garrison

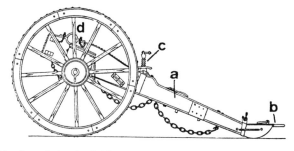

16. *9 and 6 pdr field carriage, 1845: a, trail; b, trail eye; c, elevating gear; d, capsquare.*

standing carriages and were usually mounted on a traversing slide carriage. This allowed the gun to be trained quickly on a target either to the left or to the right and, when fired, the slide checked the recoil, a necessity in the confined casemates of a fort. This type of carriage was kept in service for many years, and even when obsolete at the end of the nineteenth century still continued in use for training purposes. It can be seen in the photograph of the Scottish Artillery at gun practice taken in 1898 (plate 7).

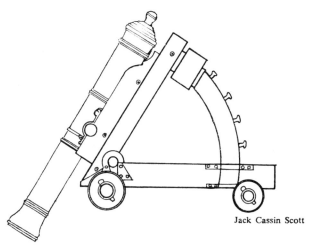

Jack Cassin Scott

17. *Lieutenant Koehler's Gibraltar depression carriage.*

When guns, either on field or garrison carriages, were positioned in one place for some time, as they would be during a siege, they were mounted on platforms made from timber, which prevented the continuous firing of the gun from driving the wheels or trucks into the ground. Such a platform is well shown in the reconstruction of a siege position at the Rotunda Museum, Woolwich (plate 4) and the photograph taken in the 1890s of the Royal Devon Volunteer Artillery at gun drill (plate 3).

The skilful use of both field and horse artillery was a decisive factor at the battle of Waterloo, where the artillery

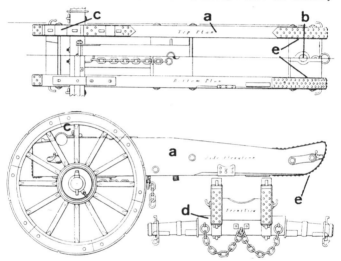

18. *18 pdr bracket carriage: a, cheeks; b, trail eye; c, capsquare; d, axletree; e, metal bands.*

stuck to their guns under the greatest difficulties and danger from the repeated attacks of the French cavalry. Only at the last moment did the gunners take refuge inside the British infantry squares but they returned to man their guns when the cavalry had passed. The Duke of Wellington spoke bitterly about the artillery, that they had disobeyed his orders to remain in the squares and had removed their guns and equipment from them. One reason for the bad feeling between the Duke and the gunners was that the gunners came under the control of the Board of Ordnance and the Master General and not of the Commander-in-Chief, Wellington. This situation was somewhat relieved when the Duke became Master General, a post he held from 1819 until 1827.

The coming of peace in 1816 saw the disbandment of a large part of the land forces including the Royal Artillery, who were considerably reduced in men and equipment. In the years that followed hardly any new equipment was added, except in the early 1820s when the deficiency created by the removal of a number of large calibre brass guns and howitzers was made good by the introduction of an 8 and 10 in. iron howitzer, which was most unusual in carriage, design. It had a combination of the old double-bracket carriage (fig. 18) and the block trail carriage (fig. 14). The top part of the carriage, where the barrel was fitted, was a double-bracket carriage, but the bottom half had a short perch or block trail (fig. 19). This block trail, which was not strong enough to support the weight of the rest of the carriage if it rested on the ground, was supported by two small iron wheels or trucks which fitted into the rear of the double-bracket part of the carriage, the block trail being purely a means to attach the gun to a limber. The ammunition box was carried on the trail, as shown in fig. 19. A novel feature was the fitting to the front of the carriage of levers which could be lashed down with rope so that they pressed on the nave of the wheel and acted as a brake.

The main improvements made between Waterloo and the Crimean War were in the field of fuses and firing methods. These are dealt with in Chapter 6.

Rifled ordnance had been experimented with for many years and tests were carried out at Woolwich in 1790 by Joseph Manton, the famous London gunmaker. Rifling had proved itself in small arms and the rifle regiments of the British Army

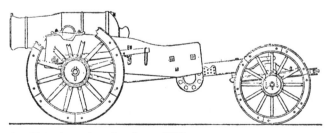

19. 10 in. iron siege howitzer, 1830.

had been equipped with the Baker rifle in 1800. After Manton's experiments and a disagreement between him and the Board of Ordnance, the idea of rifled cannon was dropped and there is no mention of any experiments with rifling in the records of the Ordnance Select Committee from 1790 until 1836.

In 1851 Charles Lancaster, another London gunmaker, who invented Lancaster's oval-bore rifling used on a carbine for the Royal Engineers, experimented with rifled ordnance. The oval-bore system was considered to have many advantages over the conventional bore and a 68 pdr Lancaster gun was approved for use by the artillery.

The artillery entered the Crimean War with almost the same equipment as that used at Waterloo, except for the Lancaster gun. During the bombardment of Sebastopol complaints were received about the poor quality of some of the ammunition, especially the carcasses or incendiary shells. However, on investigation it does not seem surprising, as many of the shells used were made between 1809 and 1814!

During the first winter of the war the complete breakdown in supplies led to the abolition of the Board of Ordnance, who had for hundreds of years been responsible for the supply of artillery, small arms and other equipment to the British soldier. The siege of Sebastopol saw also the condemnation of the oval-bore 68 pdr rifled Lancaster gun as unsafe and untrue. Over 250,000 rounds of ammunition were fired by the siege train, 32 pdrs accounting for 65,379 rounds, 8 in. guns for 64,250, as compared to the Lancasters which fired only 1,542 rounds. The siege train and naval brigade possessed the following pieces:

Mortars: 13, 10, 8 and 5½ in.
Guns: 68 pdr, 32 pdr, 10 in., 8 in., and the 68 pdr Lancaster.

During the period from 1770 to 1855, when the Board of Ordnance was abolished, there were the following Masters General:

1772–82	George, Viscount Townsend
1782–3	Charles, Duke of Richmond
1783	George, Viscount Townsend
1784–95	Charles, Duke of Richmond
1795–1801	Charles, Marquis Cornwallis
1801–6	John, Earl of Chatham
1806–7	Lord Moira
1807–10	John, Earl of Chatham

1810–18	Henry, Earl Mulgrave
1819–27	Arthur, Duke of Wellington
1827–8	Henry, Marquis of Anglesey
1828–30	Viscount Beresford
1830–4	Sir James Kempt
1834–5	Sir George Murray
1835–41	Richard, Lord Vivian
1841–6	Sir George Murray
1846–52	Henry, Marquis of Anglesey
1852	Henry, Viscount Hardinge
1852–5	Fitzroy, Lord Raglan

Between 1770 and 1855, the following types of ordnance were in use with the artillery:

Guns: Brass—42 pdr, 32 pdr, 24 pdr, 18 pdr (obsolete by 1816), 12 pdr, 9 pdr, 6 pdr, 3 pdr, and 1 pdr (obsolete by 1800).
Iron—68 pdr, 12 in., 10 in., and 8 in. shell gun (all introduced after 1820), 42 pdr, 32 pdr, 24 pdr, 18 pdr, 12 pdr, 9 pdr, 6 pdr.

Mortars: These were the same as in 1770 (see page 16) except that a new version of the 10.8 and $5\frac{1}{2}$ in. were introduced in the 1840s.

Howitzers: Brass—10 in., 8 in., (obsolete by 1813), light $5\frac{1}{2}$ in. (obsolete by 1830s), heavy $5\frac{1}{2}$ in., 4 2/5 in. (relegated to mountain artillery after 1840). 12 pdr, 24 pdr and 32 pdr (introduced in the 1820s and 1840s respectively), Iron—10 in., 8 in., (introduced about 1820), $5\frac{1}{2}$ in. (introduced in 1840).

1855-1899

'The mode of making large iron guns, by casting them in solid masses (as at present), is highly objectionable, and is certainly not suitable for bearing the full strain of a rifled gun. It is well known that if iron be cast in large masses, great irregularities will be produced in the metal during cooling; and, beyond a certain limit, little or no increase in strength is gained by increasing its thickness. Improved modes of construction can, however, be adopted, which will admit of the gun being loaded at the breech when required, and will give all the strength necessary for as large a charge of powder as can be consumed with the projectile intended for the piece.'

This was how Joseph Whitworth F.R.S., the well-known

scientist and inventor, described the construction of his new type of rifled ordnance. At the same time a civil engineer, William Armstrong, was working on a built-up gun which was rifled and a breech-loader. Rifling, which by this time had been adopted for the soldiers' Enfield rifle, was the name given to grooves inside the barrel which imparted spin to the projectile. Because the projectile fitted tightly into the barrel, on firing it spun as it went up the barrel and continued to do so in its flight. The tightness of the fit and the stabilizing effect of the spin gave the projectile greater range than a loose fitting one. With the introduction of elongated projectiles rifling stopped them from turning end over end in flight (see Chapter 6).

Whitworth's gun had a hexagonal bore and projectile, with a breech that was closed by a screw that had to be completely removed to insert the projectile and the charge. Both the hexagonal bore and the difficult loading method gave rise to complications and the Whitworth gun was never adopted.

The Armstrong gun instead of using a cast barrel also employed a built-up barrel. Armstrong's barrel was built up by shrinking thin iron coils over an inner tube (see fig. 20). His rifling was a number of shallow grooves with a special shell that was coated with lead and made slightly larger than the bore. The lead coating acted as a gas seal and driving band. The breech was closed by a wedge very like that of the early breech-loaders (figs. 3 and 4). The projectile and charge were loaded through the bored-out breech screw, and the vent piece, which not only acted as a wedge, but also was bored out to allow ignition of the charge (fig. 20), was dropped into place and held tight by the breech screw which was wound up tight against the vent piece.

Armstrong started his experiments in 1854 and delivered

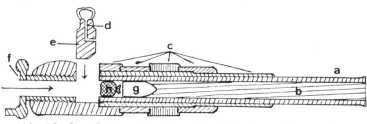

20. *The barrel of an Armstrong breech-loader: a, inner tube; b, bore with rifling; c, various coils and jackets; d, vent; e, vent piece; f, breech screw; g, shell; h, charge.*

his first 3 pdr gun to the Ordnance in 1855. His guns were tried, with the result that in 1859 orders were placed with Armstrong for field guns of various calibres. 12 pdrs were ordered as field artillery.

A large number of European countries had also been experimenting with rifling and breech-loaders with the result that Austria, Holland, Italy, Spain, and Sweden adopted rifled muzzle-loaders with studded shells that fitted the shallow rifled grooves of the barrel. Some countries, amongst them Prussia, adopted the rifled breech-loader. The French, however, retained their old brass ordnance but rifled them.

Whitworth continued his experiments, but only with rifling, so that his guns were of the muzzle-loading type. This resulted in a trial between Armstrong and Whitworth held in 1863. The Committee reported in August 1865 that,

The many-grooved system of rifling, with its lead-coated projectiles and complicated breech-loading arrangements, entailing the use of tin cups and lubricators, is far inferior, to the general purpose of war, to both of the muzzle-loading systems, and has the disadvantage of being more expensive, both in original cost and in ammunition.

That muzzle-loading guns can be loaded and worked with perfect ease and abundant rapidity.

The change to breech-loaders had by no means been completed by the time the trials were held. In March 1866 the War Office instructed the Ordnance Select Committee to draw up a schedule of muzzle-loading guns to meet the requirements of the service. The report was then put before a Committee of Superior Officers of the Royal Artillery to consider the breech-loading or muzzle-loading systems. On 4th December 1866 they unanimously reported,

That the balance of advantage is in favour of muzzle-loading field guns, and that they should be manufactured hereafter.

They further recommended

For field batteries—a gun of not less than 3 in. calibre, length of bore not to exceed 72 ins. and weight not to exceed 8 cwt; weight of projectile to be 12 lbs, or thereabouts.

They recommended that the guns of field and horse artillery should be of the same calibre.

In January 1867 the War Office decided that for the moment no change should take place, but this decision was reversed in 1870 when the return was made to the muzzle-loading gun. In 1871 the 16 pdr rifled muzzle-loader (R.M.L.)

gun with wrought iron carriage, instead of a wood one, was adopted for field artillery. The decision was thought at the time to be a wise one, especially when more than 200 Krupp breech-loaders had failed in the hands of Prussian gunners in the Austro-Prussian War of 1866.

The new muzzle-loaders employed Armstrong's system of built-up barrels with rifling that had three grooves. This was known as the 'Woolwich' system, having been invented at the Royal Gun Factory. This system employed a studded shell (fig. 21b) which fitted into the grooves of the barrel. By 1867 the Royal Gun Factory had replaced Armstrong's costly method of barrel making with their own. Instead of the many coils shrunken on to the inner core the Factory used fewer coils, but much thicker ones, so that less were needed.

During the return to muzzle-loaders the field artillery were equipped with 9 pdrs, 16 pdrs and a 13 pdr (introduced at the end of the period). For siege use there was the 25 pdr, 40 pdr, and 6.6 in. guns. Medium guns were classed as 64 pdrs and 7 in. guns together with all converted smooth-bore ordnance. In the class of heavy guns there were the 7 in. (6½ tons) up to 17.72 inch guns. The largest gun made at the Royal Gun Factory was the 68 ton gun of 13.5 ins., the larger guns of 17.72 ins. being made by Armstrong at his Elswick works.

In 1878 a new gun was introduced for both field and horse artillery. This was the 13 pdr rifled muzzle-loader (R.M.L.) which employed an elevating arc rather than an elevating screw which was normally used on field guns (fig. 22). Perhaps the most universal of all field guns at this time was the 9 pdr, which was used throughout the Empire, various marks and versions being made for India and for naval service. At the beginning the 9 pdr had weighed only 8 cwt but in 1874 a new lighter version weighing a mere 6 cwt was produced. Although obsolete by 1895 this gun continued in use as a training weapon in Canada and other parts of the Empire. The

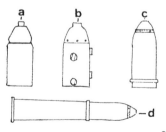

21. *Shells: a, lead-coated Armstrong shell; b, studded shell, Woolwich system; c, shell with driving band; d, fixed ammunition.*

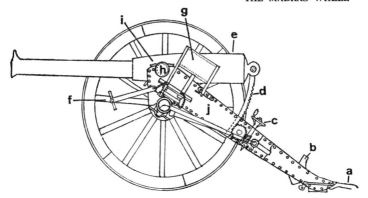

22. *13 pdr R.M.L. (rifled muzzle-loader), 1879, with its unique elevating arc (d): a, trail eye; b, socket for hand-spike; c, elevating gear; d, elevating arc; e, vent; f, foot rest; g, axletree seat; h, trunnion; i, capsquare; j, trail.*

land-service carriage used on the 9 pdr and with slight modifications on other calibre guns was almost entirely made of iron and steel and had axletree seats, adopted in 1870, so that two members of the detachment could ride with the gun, while the others rode on the draught horses and limbers (fig. 22 and plates 10 and 11).

The wheels of gun carriages had hardly altered since Waterloo, until in 1861 two Madras-pattern wheels were sent from India to England. After much experimenting a modified form of the Madras wheel was adopted on 16th September 1868. Although of the same overall size the Madras wheel differed in construction and material used.

By 1880 it was found that the loading of some guns had become difficult due to the lengthening of barrels. This had come about because it was discovered that a large charge of slow-burning propellant in a longer barrel gave a higher muzzle velocity than a similar charge of fast-burning powder in a gun with a shorter barrel. After having tested a 6 in. and 8 in. Armstrong breech-loader the decision was taken in 1881 to return to breech-loading ordnance.

By 1885 the field and horse artilleries had been re-equipped with a 12 pdr gun. Such a gun is shown in plate 12 with a detachment of Royal Marine Artillerymen preparing to load, while another appears in plate 11 complete with horses and limber ready to move. Note the two gunners on the axletree

27

seats and the three gunners on the limber. This plate also shows the harness with horse collar in use.

Howitzers had been made obsolete by the introduction of rifled guns but in 1896 a number of field howitzer batteries were formed and armed with 5 in. breech-loading howitzers. The siege batteries were armed with 6 in. howitzers.

A new type of gun introduced in about 1892 and called the quick fire gun or the Q.F. or Q gun, had been used in the Navy but a suitable carriage had not yet been designed that would take up the recoil for use on land. The only methods for checking recoil had been the recoil shoe and the axle spade. The former consisted of a metal wedge, worked by a hydraulic system, which was dug into the ground and, when the gun fired, took up the recoil. To aid the recoil shoe the end of the trail was fitted with a spade which also dug into the ground. This use of the recoil shoe made the gun jump after each round so that it had to be re-layed. It was not until after the Boer War that a suitable carriage with recoil control was designed.

The composition of a field battery of horse artillery in 1898 consisted of:
- 5 officers
- 8 artificers
- 2 trumpeters
- 76 gunners
- 1 store man
- 9 sergeants
- 6 corporals
- 6 bombardiers
- 59 drivers

Total personnel 172.

The armament, wagons and horses were as follows:
- 6 12 pdr breech-loaders
- 6 ammunition wagons
- 1 ammunition store wagon
- 1 store wagon
- 6 gun carriages
- 1 forge wagon
- 2 artillery wagons
- 102 draught horses and 29 riding horses.

1899-1918

At the outbreak of the Boer War in South Africa the British artillery was equipped with a variety of breech-loading ordnance. The field batteries, whose gunners rode on the gun and limber, were equipped with a 15 pdr field gun and the 5 in. howitzer. The 6 in. howitzer, as was said earlier, was reserved for the siege train.

It was soon apparent from the type of fighting that larger and heavier guns were needed than those in use, and to fill this gap a number of large-calibre naval guns were mounted on field carriages. These were mainly 4.7 in. guns. Some were removed from coastal forts in Cape Town and mounted on railway carriages, one such being a 22 ton 9.2 in. gun nicknamed Kandahar. The artillery did possess a large-calibre gun, the 5 in. position gun, but not in very great numbers.

The new form of artillery warfare used in the Boer War was that of firing at an unseen target. This required a new skill, which the gunners soon mastered. It was apparent from the reports from the war that the British guns were out of date compared to the advances made by the French and Germans, many of whose guns were being used by the Boers. In 1896 France had adopted her legendary '75' which was capable of firing 20-30 rounds a minute. The recoil was taken up by the top carriage, that is to say the carriage had a pneumatic system with return spring that allowed the barrel only to recoil and return to its normal position, rather than the carriage moving.

The sieges of Mafeking and Kimberley saw three strange pieces of ordnance pitted against the modern French and German weapons possessed by the Boers. These three guns succeeded in keeping the Boers at bay and, although they are not, strictly speaking, British artillery, they are of great interest to the historian, the wargamer and the model maker. The three guns were *Long Cecil*, named after Cecil Rhodes (plate 14), *Lord Nelson*, and the *Wolf*. The only one with a proper ordnance barrel was *Lord Nelson*, which was an 8 cwt gun of 1770. The barrel marked 'No 6 Port' shows it to be an old naval gun, hence its name. The old gun was presented to an African chief and having been used for a while in inter-tribal wars, lay buried for about 20 years before being dug up and given to the military forces at Mafeking for use during the siege.

The *Wolf*, which was the shortened version of 'the wolf that never sleeps', the name given to Baden-Powell by the natives, was a $4\frac{1}{2}$ in. howitzer made by Major Panzera from a drain pipe. The gun formed part of the central design of the famous Mafeking siege £1 note.

The third gun used at Kimberley, *Long Cecil* (plate 14),

was made entirely from scratch at the De Beers Company works. The designer, an American engineer named Labram who had never made anything like a gun in his life, started with an ingot of steel 10 ft. 6 in. long and $10\frac{1}{2}$ in. in diameter, and he forged a gun of 4.1 in. calibre, which fired a 28 lb shell. The entire gun, including carriage and wheels, took 28 days to make and during the siege fired 255 rounds. Its extreme range was recorded as 8,000 yards.

While the fighting was still going on in South Africa, Sir Henry Brackenbury, Director General of the Ordnance, was appointed to report on the rearmament of the artillery and his findings, which called for the re-equipment of the artillery with a new quick firer within three years instead of the normal ten, were adopted without reservation. Various armament companies at home were approached to design a new type of gun but were either incapable of this or already had too much work on hand. A German engineering company called Ehrhardt, who had never been armament people, produced a revolutionary design of 15 pdr which, like the French '75', had the recoil taken up by the top carriage. After only a small trial orders were placed in secret for enough guns and limbers to re-equip eighteen batteries of field artillery.

The new Ehrhardt gun for field artillery was officially adopted in June 1901, although manufacture had been authorised in 1900 (plate 15). The gun was provided with pneumatic buffers and a telescopic trail as well as having a spade to check the recoil. The new piece had a calibre of 3.1 ins. and a muzzle velocity of 12,525 feet per second. By June, the date of its official adoption, it had already been supplied to a number of field batteries.

This new weapon, however, proved to be only a stopgap, as in 1901 a committee was set up to consider the rearmament of both field and horse artillery. Specifications of the ideal gun required were drawn up and various companies were invited to submit designs which would comply with them. By 1902 various designs of gun had been submitted and were under trial. While they showed many good features, an ideal gun was not amongst them. It was therefore decided to take the best points from each and to incorporate them into one gun. The barrel would be wire-wound on the Armstrong principle while the recoil system would be that submitted by Vickers; other parts would be decided upon by the Ordnance. It was decided that the gun should be in two calibres, 13 pdr for the horse artillery (plate 16) and 18 pdr for the field batteries. By 1903 four complete batteries were ready for trial. From the beginning the 13 pdr showed itself to be a far better shooter at all ranges than the 18 pdr and it was suggested that the 13 pdr should be common to both horse and field artillery as the 18 pdr was not considered good enough to justify manufacture.

Immediately, further trials were ordered using a 14½ lb shell for the 13 pdr and a 20 lb shell for the 18 pdr with the result that the 14½ lb shell was found to be the more accurate. The committee could still not decide whether to adopt both equipments or just the 13 pdr. Experts were called in to give their views, but still no decision was reached. At last the issue was put before the Prime Minister, Balfour, who opted for having both equipments, the 13 pdr for the horse artillery, and the 18 pdr for the field artillery. This decision was to have great influence on the artillery effort of the 1914-18 war.

There was considerable delay in the placing of orders for the new guns, but public and newspaper pressure forced the Government to issue the orders on Christmas Eve 1904. It was not, however, until 1906 that seven divisions were equipped with the new 18 pdr and this only after the many preliminary problems that go with the mass production of a new weapon. One of the most sinister difficulties that arose was the report that many of the barrels were warped, but, to the disbelief of many artillery officers, the President of the Ordnance Committee stated that a shell passing through the barrel at the time of firing would momentarily straighten it. This, in fact, worked.

At the same time as the work on the 13 and 18 pdr guns, a new howitzer was being designed to replace the 5 in. weapon of Boer War vintage. In 1908 it was recommended that the newly designed 4.5 in. howitzer be adopted. Perhaps the most lasting effect of the Boer War was the need for heavy guns, which was temporarily filled by the creation of a heavy artillery brigade equipped with 4.7 in. guns. After many experiments the 60 pdr heavy gun was approved. It had a rate of fire of two rounds a minute and an extreme range of about 10,300 yards.

At the outbreak of war in 1914 a division comprised three artillery brigades, each consisting of three batteries of six 4.5 in. howitzers, three batteries of six 18 pdr guns, and a heavy battery of four 60 pdrs (plate 18). However, in October 1914 some divisions possessed only six 18 pdrs instead of the fifty-four prescribed. The shortage of ordnance and shells began to get worse and after Lord Kitchener's call to arms, the position deteriorated further. Gunners were completing their training without even seeing a gun and many obsolete 12 and 15 pdr guns were put into service as training weapons.

To meet the shortage even wooden dummy field guns were made! By May 1915 the problems seemed to have been partially solved but an even graver problem had arisen, that of shortage of ammunition.

At the first battle of Ypres field batteries had been rationed to twenty rounds per gun per day, later reduced to ten and finally to two. The 4.5 in. howitzers were rationed to two rounds and the 6 in. howitzers to six rounds. This state of affairs led to the formation of the Ministry of Munitions under Lloyd George which set out to solve this shortage.

There was a demand from the British Expeditionary Force in France for more heavy guns and this demand was filled by using Mk 7 naval guns on field carriages. As the war progressed and became trench-bound and stagnant the following heavy guns were used:

15 in. howitzer, approved in 1914 and ideal for bombardment work.

12 in. gun which fired a shell weighing 850 lbs.

12in. howitzer

13.5 in. gun

9.2 in. howitzer

8 in. Mk VI howitzer with an extreme range of about 10,000 yards.

By June 1915 the 6 in. howitzer (30 cwt) was replaced by one of the same calibre weighing only 26 cwt and with a range of about 10,000 yards. Although the 5 in. howitzer of the Boer War was obsolete, having being replaced by the 4.5 in. howitzer, a number were used in Gallipoli in 1915. In Africa use was made of 12 pdr naval field guns.

The decision taken by Balfour in 1904 showed itself to be a wise one, for during the war of 1914-18 the 18 pdr field gun of the Royal Artillery fired nearly 100 million rounds, compared with the 13 pdr which fired only $1\frac{1}{2}$ million.

3. HORSE, MOUNTAIN AND COASTAL ARTILLERY

Horse artillery up to 1855

Although there had been various attempts at mobile artillery from the early 1700s, it was not until the formation of the Royal Horse Artillery in 1793 that Britain possessed an artillery which could keep up with the cavalry. Gustav

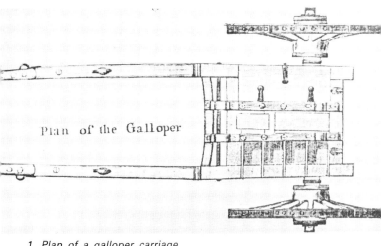

1. Plan of a galloper carriage.

2. Royal Artillerymen about to fire a 9 pdr gun; note the position of the gunner with rammer and the firer who has just ignited the gun with a portfire.

3. 2nd Devon Volunteer Artillery at gun practice with 64 pdrs on wooden garrison carriages.

4. 32 pdr on an iron garrison carriage.

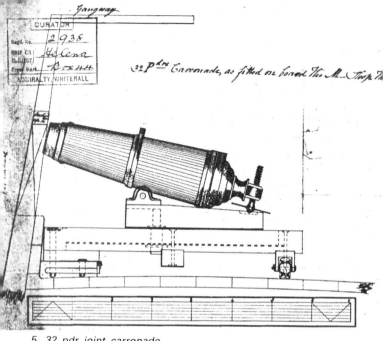

Gangway

32 Pdr Carronade, as fitted on board His M. Ship T

5. 32 pdr joint carronade.

6. The Museum of Artillery in the Rotunda, Woolwich, London.

7. *Scottish artillery at garrison gun drill, 1897.*

8. *24 pdr howitzer on field carriage, 1864.*

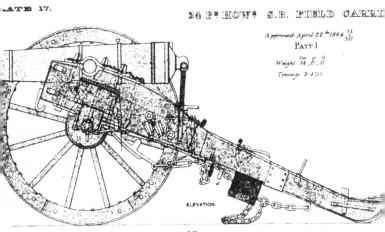

ATE 17.

24 P^{r} HOW^r S.B. FIELD CARRI

Approved April 27th 1864 $\frac{71}{310}$

PATT 1

Weight 14 $_{,,}$ 6 $_{,,}$ 0

Tonnage 3·425

ELEVATION.

9. Early nineteenth-century 12 pdrs on Prince of Wales bastion at Fort George, Inverness.

10. Plan and elevation of a 9 pdr bronze gun (for India), 1870.

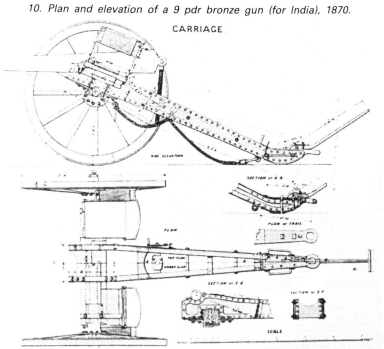

CARRIAGE

SIDE ELEVATION

SECTION at A B

PLAN

PLAN of TRAIL

TOP PLAN
LIMBER PLAN

SECTION at C D

SECTION at E F

SCALE

11. *A field gun team of the Queensland Permanent Artillery. Note the gunners on the limber seats and axletree seats.*

12. *Royal Marine artillerymen at drill on a 12 pdr breech-loader.*

13. *A French howitzer.*

14. *'Long Cecil', the Kimberley siege gun.*

15. 15 pdr Ehrhardt gun adopted for field batteries in 1901.

16. 13 pdr gun, the armament of the Royal Horse Artillery during the First World War and still in use with King's Troop RHA today.

17. Early eighteenth-century 12 pdr gun on renewed field carriage at Tilbury Fort, Essex.

18. 60 pdr gun firing, France, 1916.

19. Early nineteenth-century 13 inch mortar at Fort George, Inverness.

20. Six 13 inch mortars of Crimean War date (1854-6) at Tilbury Fort, Essex.

21. *English cast iron 64 pdr RML Palliser gun of 1871 on traversing siege carriage, for coastal defence, at Pendennis Castle, Cornwall.*

22. *13 pdr gun mounted on a lorry as an anti-aircraft gun, 1915.*

23. Indian mountain artillery battery assembling a 2.5 inch 'screw gun'.

24. Muzzle-loaders in the gunnery shed at Whale Island in the same position as they would be on a broadside ship, 1890.

25. Early rifled gun, an Armstrong 64 pdr of 1865, on a traversing platform on the Duke of Cumberland's bastion, Fort George, Inverness.

26. Typical proving marks on a gun. They show that it is a 6 pdr cast by Bailey, Pegg and Company (of Wapping, London) in 1835 and that it weighs 4 cwt, 0 quarters, 13 lbs. The crown is a trademark. Pieces such as this were made for merchant vessels.

27. A 6 inch quick firer (QF) or Q gun with protective shield.

28. A Mk III, 9 inch RML Armstrong-Frazer gun of 12 tons, introduced in 1868, at Southsea Castle, Portsmouth.

29. A Mk VII 8 inch breech-loader Armstrong gun.

30. A heavy howitzer in action, 1916.

Adolphus, King of Sweden, is credited with being the first to employ 'horse artillery' when he used two light 4 pdrs on light carriages capable of keeping up with his cavalry.

Another use of mobile artillery was the galloper gun (plate 1), designed in 1745 but discontinued in 1747. The galloper consisted of a gun mounted on a split-trail carriage which could be pulled by one horse. The need for a form of light artillery capable of keeping up with infantry or cavalry was filled by allotting to the infantry battalions 6 pdr guns manned by gunners of the Royal Artillery. The decline in morale of the gunners, many of whom had spent years with the infantry, led to the decision to train one officer and eighteen men of each cavalry regiment and one officer and 34 men of each infantry regiment to serve the guns. By the time this order had been implemented the use of battalion guns had died out and the Royal Horse Artillery had been formed.

The Royal Horse Artillery had all gunners mounted on horses, and drivers were supplied by the Corps of Artillery Drivers, a body separate from the Artillery but responsible for bringing up the ammunition wagons and driving the carts and other carriages.

The armament of the four troops formed in 1793 was a mixture of two 12 pdrs, two 6 pdrs and two $5\frac{1}{2}$ in. howitzers, but by 1800 the 12 pdrs had been discarded. In the Peninsular War so successful were the 9 pdrs of a reserve brigade that they were added to the armament of half Wellington's horse artillery. By the time of the Waterloo campaign a number of troops had five 9 pdrs (fig. 14) and one $5\frac{1}{2}$ in. howitzer, whereas others had 6 pdrs and a howitzer. One troop had rockets attached to it and another, Bull's troop, had six heavy $5\frac{1}{2}$ in. howitzers.

The typical establishment of a troop is as follows (actually G Troop in June 1815):

5 9 pdrs, 1 $5\frac{1}{2}$ in. howitzer with 8 horses each
9 ammunition wagons with 6 horses each
1 spare wheel wagon with 6 horses
Forge, curricle cart, baggage wagon with 4 horses each.
(Total number of horses 120.)

For the rest of the detachment and other troop personnel, as well as baggage animals and officers' horses, there were a further 106 animals.

The personnel of G troop were as follows:

First captain	1 farrier
Second captain	3 shoeing smiths
3 lieutenants	2 collar makers

1 surgeon	1 wheeler
2 staff sergeants	2 trumpeters (one was a driver)
3 sergeants	80 gunners
3 corporals	84 drivers
6 bombardiers	

Total personnel 193

The troop consisted of three divisions, each with two pieces, each piece with its wagons and carriages being a sub-division.

In 1813 the twelve troops of Royal Horse Artillery had two further troops added to them. These were the First and Second Rocket Troops, who were armed exclusively with rockets. Rockets had been used for many years by the Indians and Chinese and were first employed against the British at the battle of Seringapatam in 1799 with much success. This prompted William Congreve to experiment with rockets in England, the first of which were used in 1805 and in 1806 in the attacks on Boulogne. It was soon realised that rockets could be just as useful fired on land as they could be fired from boats at land targets and the Prince Regent ordered the formation of a Rocket Corps.

In action the rocket could be fired either along the ground or from a small metal platform. At Waterloo 58 rockets were fired but, judging from contemporary accounts, most seem to have proved as dangerous to the British as to the French! The types of rockets were classed in the same manner as guns:

Heavy rockets; 8, 7 and 6 in.
Medium rockets; 42, 32 and 24 pdrs
Light rockets; 18, 12, 9 and 6 pdrs

The rocket consisted of three parts: the stick, which was in two or more sections depending on the size of the rocket; the charge; and the projectile, which was shell or case shot. On the march the gunners of a rocket troop carried the sticks in bunches which gave them the appearance of lancers. A small spearhead was supplied to each man to convert a rocket stick into a lance if required!

In 1816 one rocket troop and four gun troops were disbanded, and by 1819 there were only seven troops left. The Royal Horse Artillery was now armed with 6 and 9 pdr guns only and this was to be their armament at the outbreak of the Crimean War. When horse artillery troops were dispatched to Crimea their armament was augmented with 12 and 24 pdr howitzers.

In Crimea the horse artillery were in action on a number

of occasions and featured in the famous note from Lord Raglan that ordered the charge of the Light Brigade:

'Lord Raglan wishes the cavalry to advance rapidly to the front, follow the enemy and try to prevent the enemy carrying away the guns. Troop horse artillery may accompany. French cavalry is on your left. Immediate.'

Although the horse artillery did prepare to follow the Light Brigade, they were not able to advance very far because part of the valley was ploughed.

From Crimea a number of troops were dispatched to India to deal with the Mutiny, the first time that the Royal Horse Artillery had been in India as the East India Company had possessed its own horse artillery.

Horse artillery 1855-1899

With the coming of the breech-loader (see page 23) the Royal Horse Artillery were ordered 9 pdr Armstrong guns in place of their muzzle-loading 6 and 9 pdr guns and howitzers.

The horse gunners were just becoming accustomed to their new equipment when it was decided that the breech-loader not only was more expensive, but also had no marked superiority over the proven muzzle-loaders. In 1871 a 9 pdr rifle muzzle-loader (R.M.L.) with a wrought iron carriage was adopted by the horse artillery.

As a result of the Franco-Prussian War the emphasis was placed on greater mobility for the horse artillery, and a new 9 pdr gun of only 6 cwt, compared with the old pattern which weighed 8 cwt, was adopted in 1874. No limber or axletree seats were fitted on the 9 pdr for horse artillery as the detachment rode their own horses with the gun and limber which was pulled by a six-horse team.

In 1878 it was felt that a heavier gun was needed for the horse artillery and accordingly a 13 pdr rifled muzzle-loader (R.M.L.) (fig. 22) was adopted for both horse and field artillery. This was the gun that took part in the Zulu War of 1879 and at Maiwand during the Second Afghan War, in 1880, where two Victoria Crosses were won by the Royal Horse Artillery during the action to save the guns.

In 1881 the return was made to breech-loading guns and by 1885 the horse artillery had been re-equipped with the 12 pdr rifled breech-loader (R.B.L.) (plates 11 and 12) which was the same gun as supplied to the field artillery. In 1894 the horse artillery was supplied with a 12 pdr gun of only 6 cwt, 1 cwt less than the gun adopted in 1885. This was the gun that the horse artillery was to use throughout the Boer War.

Horse artillery 1899-1918

Although the Royal Horse Artillery was in action many times in South Africa and contributed gallantly to the war, the emphasis was on heavier artillery. Because of Boer tactics it was seldom that the artillery was in close action with an enemy it could see. In 1900 the field batteries had been ordered a new gun, the Ehrhardt (plate 15), which had a newly designed carriage that took up the recoil (see page 30) but the horse artillery still had its 12 pdrs.

It was not until 1902 that a new gun was considered for the horse artillery and this, the 13 pdr (plate 16), was not ordered until 1904 after many trials and troubles (see page 31).

At the outbreak of the 1914-18 war there were 27 batteries of horse artillery. Although the first few months of the war saw the horse artillery in its true mobile role, the trench war that soon evolved saw them either in the line with the field gunners or sometimes with the cavalry, anticipating the 'break through' that never came, and, even if it had come, it is doubtful whether over the muddy waste land pitted with shell holes they could have achieved their expected mobility.

The beginning of the war, during the retreat, was the time when the horse artillery was in its element. This is perhaps typified by the stand of L Battery at Nery, where despite heavy shelling from the Germans the last remaining gun was worked by the only two survivors until the last round of ammunition.

In the desert in Palestine and by the Persian Gulf the horse artillery found more scope and, in keeping up with the advancing cavalry, were able to fulfil their true role.

A new form of warfare that did not really affect the horse artillery, but which is included here because it was their 13 pdr that was first used to combat it, was war from the air. No specially designed guns existed at the outbreak of war, and a number of 13 pdrs were mounted on lorries as mobile batteries (plate 22). A new anti-aircraft gun was designed in 1914, which was a 3 in. gun, but the 13 pdrs mounted on their lorries continued in use.

As is usual at the ending of a war reductions were made in the armed forces. In the horse artillery the 27 batteries were reduced to fifteen by 1920.

The Royal Horse Artillery, as the King's Troop, still continues to use the 13 pdr horse-drawn gun for ceremonial occasions (plate 16), with the men dressed in the uniforms of the pre-1914 era.

Mountain artillery

The mountainous country of the Pyrenees was responsible for the formation of another type of light artillery, the mountain artillery, formed in 1813 and equipped with brass 3 pdr guns which were dismantled, the various components being carried on the backs of three mules. The mountain artillery performed good work in driving the French out of Spain, but was perhaps short-sightedly disbanded at the end of the war.

The artillery of the Honourable East India Company were not slow to learn from this new use of light artillery and formed a camel battery which was suitable for the hilly

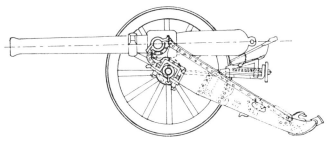

23. Elevation of a 2.5 in. mountain gun.

country and for the sandy wastelands in which the Mahratta War of 1819 was fought. Although used by the British again in Spain in 1836, mountain artillery was more used in India.

The mountain artillery was formed as a permanent force in 1849 and was part of the Punjab Frontier Force which defended the North West Frontier. The batteries were armed with 3 pdr guns and 4.4 in. howitzers, both of smooth bore. When dismantled each gun was carried by three mules, the barrel on one, the carriage on one, and the wheels on the third.

When the artillery of the East India Company was absorbed into the British Army the Indian Army kept only the four mountain batteries. In the 1860s a steel-barrelled 7 pdr rifled gun was adopted by the mountain batteries. This weapon had a metal carriage and weighed 200 lbs with an extreme range of 3,000 yards. In 1876 the introduction of slow-burning powder required a longer barrel and the 2.5 in. 'screw gun', immortalised by Kipling in his famous poem 'Screw Guns', was adopted. The novel feature of the 2.5 in. mountain gun (plate 23 and fig. 23) was that the barrel was in two halves.

Before going into action the halves had to be screwed together (plate 23), and hence its name.

The 2.5 in. gun was in use until the end of the nineteenth century. Even though cordite, a smokeless propellant, had been introduced in the 1880s, the mountain gun still fired gunpowder. The dense clouds of smoke produced on firing not only gave away the gun's position but also allowed the enemy time to take cover.

In 1902 a new breech-loading mountain gun was introduced. This was the 10 pdr mounted on a metal carriage. In 1911 the gun was modified by fitting a new carriage with hydraulic recoil mechanism. Hitherto the recoil of the mountain guns had been checked by drag ropes lashed from the wheels to the trail. This gun was renamed the 2.75 in. mountain gun.

Because of the nature of the country where these guns were used there were always demands for a mountain howitzer, but not until the beginning of the 1914-18 war was one adopted. This was the 3.7 in. Q.F. (quick firer) howitzer, which had a range of 6,000 yards. Dismantled it was carried by eight mules and continued in service for many years, including the 1939-45 war, when it was used against the Japanese in the jungle and mountainous country of Burma.

Coastal artillery

From the early days of artillery it was imperative to protect coastal and other fortifiations with large-calibre guns that could engage the enemy before he was able to get close enough to use his own guns. Because of this the guns used in coastal defence were large and heavy and mounted on heavy wooden carriages (plate 3). In times of peace the wooden carriages were replaced with iron ones (plate 4). In forts the guns were either mounted on the wooden standing carriage or on traversing carriages which gave greater scope to the gunners in quickly bringing the gun to bear (plate 7). The inclined carriage also took up the recoil and allowed the gunners to load quickly before the gun was slid down into the firing position. This type of carriage was kept in use for many years and was later replaced by one of similar principle, but in metal (plate 7).

Carronades (see Chapter 4) were also used in coastal defences because of their easy handling in confined spaces and their ability to fire a large projectile. They were mounted on wooden block trail carriages or on iron carriages as well as on the slide carriages used by the Navy (plate 5).

With the coming of steel guns and rifled barrels with elongated shells, a number of the old smooth-bore guns of the coastal artillery were bored out and fitted with wrought iron rifled tubes. Smooth-bore 32 pdr and 8 in. guns were converted to 64 pdrs, and 68 pdr smooth-bores became rifled 80 pdrs. These guns remained for years the sole weapons used in coastal defence, many in fact surviving as late as 1904, when the last muzzle-loader was removed from service. In 1877 the armament of the coastal defences was so varied that any form of serious training must have been nearly impossible. There were the following pieces in service:

Rifled muzzle-loaders (R.M.L.); 12 in., 10 in., 9 in., 64 pdr and 80 pdr

Rifled breech-loaders (R.B.L.); 7 in., 40 pdr, 20 pdr, 9 pdr, 6 pdr

Smooth-bore guns; 8 in., 68 pdr, 32 pdr, 24 pdr, 18 pdr, 24 pdr howitzer

Mortars; 13 in., 10 in., 8 in.

It was soon realised that the heavy slow-firing guns were inadequate against a quick attack and landing and so a large number of breech-loading guns were allocated to the coastal defences.

Another form of carriage for coastal and naval artillery was the Moncrieff disappearing carriage. This novel system enabled the breech-loading gun to be loaded in safety and under cover, without presenting a target to the enemy at the same time. Once loaded it was raised to the firing position, fired, and the barrel returned to its protected loading position. Experiments were tried with this type of carriage which included its being raised for a certain time while a ship at sea tried to score a hit. During the experiment, at Portland Bill in 1885, H.M.S. *Hercules* fired over 100 rounds at a wooden model of this gun without scoring one hit.

By the end of the century a number of heavier guns had been placed in the coastal defences, notably the 10 in. (29 ton) and the 8 in. (15 ton) guns. By 1910 the 9.2 in. breech-loader, the 6 in. breech-loader and the 12 pdr Q.F. (quick firer) (plate 26) were the standard coastal artillery in the British service.

4. NAVAL ARTILLERY

1650-1770

In the Navy the guns had hardly altered since the time of Henry VIII because it was thought that the pieces in use would not only be more liable to be lost if the ship was sunk but also would never have to fire in any engagement for as long as those employed on land. England was not alone in this thinking; France designated all its old cast iron guns to the Marine and continued to do so until about 1760.

Although in theory the guns of each calibre were of the same size this was seldom the case, especially in the Royal Navy, where it was necessary to fit the guns into certain parts of a ship. There were therefore various lengths of gun in each calibre. The carriages were plain wooden beds on to which the barrel was strapped. In early forms no trucks, small wheels, were used but they were later adopted.

Another form of artillery, if it can be called that, introduced at the end of the seventeenth century was the bomb vessel. The English first used bomb vessels in the attack on St. Malo in 1693. The bomb vessels, or 'H. M. Bombs' as they became known, were specially constructed ships of shallow draft with mortars mounted on board. With the combination of shallow draft and mortars the ships were able to get in close to bombard coastal towns and defences. At the same attack on St. Malo another form of naval artillery was employed. This was the fire ship. These were merchant vessels, purchased, converted, and fitted out as floating incendiary bombs, which, if the tide and current were right, could be set loose amongst an enemy fleet or directed at an enemy harbour or port.

By 1750 the carriages used at sea had become the standardised standing carriage (fig. 15). In the Navy wooden trucks were used instead of the iron ones employed on land, as the metal wheels would be liable not only to damage the decks but also, if hit by an enemy shot, to break up, showering dangerous splinters.

At sea iron ordnance was used, except for certain mortars and the guns of the *Royal George,* which were brass. The same types of iron ordnance were used at sea as on land (see page 16) and the following mortars were in use in 1770:

Brass; 13 in. and 10 in.
Iron; 13 in. and 10 in.

1770-1855

Throughout this period of advance in the construction of land artillery, very little seems to have been done at sea apart from supplying the new patterns of guns. The carriages altered only in small details, remaining the same as in the 1750s. All warships were still 'wood wall' ships, the guns were still arranged in lines on gun decks, and there was also no change in the loading and firing methods. In a ship of the line each gun had a crew according to its size, one man for every 500 lbs weight. Each gun was positioned on the deck opposite the gun port through which it was to fire. The gun was anchored to the side of the ship by a breeching rope that passed through the eye on the cascable of the barrel (fig. 10h) and was fixed to the side of the ship. Besides this rope, which was to take up the recoil, there was a set on each side, the left side tackle for training left and the right side tackle for training right. Immediately to the back of the carriage was another training tackle for keeping the gun straight. The guns were loaded in the same way as on land with the charge and projectile being rammed well home. In confined spaces a rammer with a rope shaft was used (fig. 13d) as opposed to one with a wooden shaft (fig. 13c). Once primed and ready the rear train tackle was released and the gun was run out with the side men holding the side tackle until the command came to fire, at which they dropped it. Back in its recoiled position the operation was then repeated.

A most important new weapon was added to the Navy's armament in 1779, this being the carronade (plate 5), made by the Carron Company of Falkirk in Scotland, and designed by Mr. Gascoine, the works manager, and General Robert Melville, an artillery officer who for a number of years had been experimenting with shorter and lighter guns. The guns appear to have been called 'gasconades', but the name was soon changed to 'carronade'. Trials were held in August 1779 and the carronade was adopted. Besides this type of ordnance the Carron Company also produced cast iron guns of various calibres. Such was the high quality of their work that the Duke of Wellington used to ask for their guns by name.

The main use of carronades was at sea and in garrisons, where their short barrels and ease of loading made them ideal weapons, as they could be fitted into confined spaces. The barrel was mounted on a slide carriage for use at sea (plate 5) by means of a loop cast underneath the barrel, and not by trunnions as was usual with other pieces of ordnance.

Although their range was not as great as that of a gun carronades were used for short-range work and their ability to fire a shot of large calibre out of all proportion to the size of barrel earned them the nickname, the 'smasher'. Carronades continued to be made throughout this period but the advent of built-up steel guns of larger size and greater range rendered them obsolete and they were withdrawn from service in about 1860, none having been made since 1852.

During their lifetime the following calibres of carronade were made:

68 pdr, 42 pdr, 32 pdr, 24 pdr, 18 pdr, 12 pdr, and 6pdr.

All of these were in use with variations in length in about 1840, except the 6 pdr, which does not seem to have been in use after 1811.

1855-1899

The development in the breech-loader and the built-up gun has already been described in Chapter 2 and the Royal Navy was soon equipping ships with guns on the built-up principle (plate 24), using the breech-loaders for field and boat guns. The Admiralty also discussed the merits of breech-loader and muzzle-loader and in December 1865 wrote to the War Office stating that they were 'averse to the perpetuation of the breech-loading system for field and boat guns in the Royal Navy'.

By 1880 it was discovered that the lengths of some of the muzzle-loading guns in the Royal Navy made loading a problem. Guns were mounted in two ways in the Navy, depending on the type of ship. There were broadside ships, with the guns arranged along the gun decks as in Nelson's day, and turret ships, which had the guns fitted in turrets on the decks. The main problem was that there had to be enough space in turrets and on gun decks to 'run in' the guns for loading. In certain cases specially protected positions had to be built outside turrets to protect the loaders!

In 1881 Armstrong invited the War Office and the Admiralty to test a 6 in. and 8 in. breech-loader, and soon the decision was taken to return to breech-loaders, with the Navy now using them on the larger-calibre guns.

In 1892 cordite was introduced, a smokeless propellant which took its name from its appearance. An advantage of cordite was that a lesser weight of it was required to propel a projectile (see list of naval guns in 1898, page 59). Great changes had been made in the guns of the Royal Navy and turret ships were now considered far superior to broadside ships. As a large number of the older ships were still armed with muzzle-loaders a programme of re-equipping with breech-loaders was undertaken for those vessels considered fairly modern. A new type of gun had been introduced in about 1892 to speed up the rate of fire. This was the Q.F. (quick firer) or Q gun (plate 27), which fired fixed ammunition (see Chapter 6).

By 1898 the many other small-calibre breech-loaders had been replaced by Q guns or had been converted. The Moncrieff disappearing carriage was not considered suitable for the Navy as turrets were more practical and could house more than one gun.

In 1898 the following types of gun were in use in the Royal Navy:

gun	length	bore	weight of shell	weight of charge
111 ton[1]	43 ft.	16¼ ins.	1,800 lbs.	960 lbs.
67 ton[2]	36 ft.	13½ ins.	1,250 lbs.	630 lbs. cordite
46 ton[3]	37 ft.	12 ins.	850 lbs.	167½ lbs. cordite
29 ton[4]	28 ft.	10 ins.	500 lbs.	252 lbs.
22 ton	25 ft. 9 ins.	9.2 ins.	380 lbs.	166 lbs.
15 ton[5]	21 ft. 2 ins.	8 ins.	210 lbs.	188 lbs.
6 in. (7 tons)[6]	17 ft.	6 ins.	100 lbs.	13¼ lbs. cordite

Q guns: 4.7 in., 4 in., 3 in. (referred to as 12 pdr) and 6 pdr with 3 pdr Hotchkiss and Nordenfelt guns.
Breech-loaders not Q guns were 6 in., 4in., 20 pdr and 12 pdr.

1. Only the battleships *Sans Pareil*, which had two of these mounted in one turret, and *Benbow*, which had one in each of two turrets, had these guns. The *Victoria* also had them but was lost in 1893. This gun was considered obsolete.
2. This gun was on fourteen battleships but was obsolete.
3. The 45 ton (12 in.) gun was on only five battleships and was superseded by the 46 ton gun.
4. This gun was on four new battleships and replaced certain muzzle-loaders in old turret ships.
5. A few 15 ton guns were in service. The *Bellerophon*, an old broadside ship, and two Indian defence turret ships had them.
6. A Q or quick firing gun. See plate 27.

1899-1918

By the time the 1914-18 war had broken out the Royal Navy had a formidable array of large-calibre guns, and one of their ships, H.M.S. *Agincourt*, carried the most heavy guns of any ship in the world, with fourteen 12 in. guns mounted in seven turrets. A super dreadnought of this period had eight 15 in. guns, twelve 6 in. guns and twelve 12 pdrs mounted in various turrets and positions. A new type of ship that was to appear during this war was the Q ship (not to be confused with the Q gun). This was a disguised merchant vessel which at a moment's notice could convert itself into a formidable fighting ship. They were used against submarines which attacked merchant ships. The Q ship, on sighting the submarine (they usually surfaced to destroy the merchant ships by gunfire, as they saved torpedoes for warships), would stage an 'abandon ship', leaving men aboard to man the guns. As soon as the submarine closed in the white ensign would be hoisted and the Q ship opened fire. Their armament consisted of two 4 in. guns and two 12 pdr Q.F. guns.

During the war the Royal Navy had the following guns:
15 in., 13.5 in., 12 in., 9.2 in., 7.5 in., 6 in., 5 in., 4.7 in., 4.25 in., 12 pdr, 6 pdr, 3 pdr, $1\frac{1}{4}$ pdr, and 1 pdr.

5. MANUFACTURE

The manufacture of cast ordnance was a lengthy procedure. A model of the barrel to be cast had to be made and this was done by building up the shape from a mixture of clay and horse dung on a spindle. Once the required thickness of layers of rope and the mixture had been reached, the model was shaped with a template cut to the contours of the barrel. Blocks of wood were hammered in to form the trunnions and the model was dried over a fire. It was at this stage that any decoration, such as the crest or cipher of the reigning monarch and the coronet and cipher of the Master General of the Ordnance, was put on the barrel.

From the dried model the mould was made of the same composition of clay and horse dung. Once this mould had been made to the correct thickness to withstand the force of the molten metal it was dried and banded with metal hoops. The model was then eliminated by lighting a fire inside which destroyed the original. The mould was dried and was then ready for the pour. Before the early part of the eighteenth century the bore was also cast in the piece.

As soon as the barrel was cool, which could take a number of days, the bands and mould were removed and the barrel was drawn out and cleaned by hand, the bore being cleaned and rectified if necessary. This method survived until 1739 when barrels were cast solid and bored afterwards. This method was not immediately adopted by all founders, but it was employed at Woolwich. It was not until 1775 that the Board of Ordnance stated that it would only receive barrels bored from the solid.

The casting of barrels followed the same procedure until

the introduction of the Armstrong barrel built up with coils on an inner tube (fig. 20). This new type of barrel was used both on muzzle- and breech-loading ordnance. By 1867 the Royal Gun Factory at Woolwich had replaced the Armstrong barrel of many small coils with one of their own with fewer, larger coils, which was easier and cheaper to make. Towards the end of the century this system was replaced by the wire-wound method, which involved the winding of thin wire over an inner tube before shrinking on the outer jacket. In principle this is the system still in use today.

Although steel barrels had replaced cast barrels in the British service the last cast bronze gun was approved in 1870 and continued in the hands of the Indian Artillery until the turn the of the century (plate 10).

6. PROJECTILES AND FUSES

Until the advent of rifled artillery and breech-loaders the projectiles consisted of round shot, canister or case shot, langridge shot, grapeshot, bar shot, chain shot and shells. This does not include the incendiary carcasses used to set fire to enemy towns and fortifications.

The various forms of these projectiles are shown in figs. 13 and 24. Each type had a specific job; round shot against walls and earth fortifications; canister, case, langridge and grape against formations of troops; bar and chain shot were employed by the Navy to cut down the rigging of an enemy ship; and shells were used against towns, fortifications and formations of troops.

Perhaps one of the most far reaching inventions was that in 1784 of spherical case shot or 'shrapnel', as it became called after its designer Lieutenant Henry Shrapnel (fig. 24c). It was similar in construction to the common shell but, instead of having only bursting powder inside and relying on the frag-mented case when it exploded for its destructive powers, Shrapnel filled the inside of his shell with both bursting powder and musket balls. The advantage of shrapnel was that when the shell exploded and the case burst open the balls continued in the line of flight. The small charge of bursting powder was used so that the shot did not scatter too much, and so was more effective. In 1803 this new idea was adopted by the Ordnance.

Grapeshot, which took its name from its resemblance to a bunch of grapes, was composed of a number of iron balls packed round a central column and covered in canvas. This was then 'quilted' by passing cord across each bulge in the bag (fig. 24f). Case shot was a number of balls enclosed in a tin case. The common shell, as has been stated, was a hollow round shot filled with a bursting charge.

Fixed ammunition was ammunition that had the charge and projectile in one. In the early days round shot, case shot and shells were fitted to their powder charge so that loading was made easier. To make fixed ammunition the projectile had to be fitted with a wooden base or 'sabot', which in turn was forced into a bag containing the charge and tied (fig. 24h). The advantage of the sabot was that it prevented the shot or shell from turning as it went up the bore of the barrel. The sabots were held to the shell by iron straps (fig. 24a) which were replaced with a rivet (fig. 24d) by 1850.

The advent of rifled ordnance rendered round shot and shell obsolete as now these guns fired either elongated shells coated with lead (fig. 21a), which acted as a driving band up the bore, or in Whitworth's case a hexagonal shell which fitted the hexagonal bore. The lead-coated shell was satisfactory for the Armstrong breech-loaders but was unusable in a muzzle-loader of any calibre as the shell would become distorted while being rammed down the rifled bore.

The French, who had retained their muzzle-loaders but rifled them, had adopted a rifling of shallow grooves which corresponded with studs on the shell. With the return to muzzle-loaders in 1870 the Woolwich system was adopted. This employed a three-groove rifling with a shell with corresponding studs (see fig. 21b). It was found, however, that the windage when using studded shells caused erosion on the bore of the barrel and to combat this a *papier mâché* cup was placed between the shell and its charge. In 1878 a copper cup placed at the base of the shell replaced the *papier mâché* cup. This became known as a driving band (fig. 21c) when it was found that by fitting the copper to the shell it did the work of the studs in giving the shell its spin.

When the Q or quick firer gun was introduced the charge was in a brass case with the projectile fitted (fig. 21d), which sped the rate of fire.

The use of an explosive shell depends on the fuse and until the 1750s this was very rudimentary. From this date onwards fuses consisted of a beech-wood cone (fig. 13e) ribbed on the outside in seconds and filled with powder. Once the required length of time was known the fuse was cut off at the corresponding rib before being put into the shell.

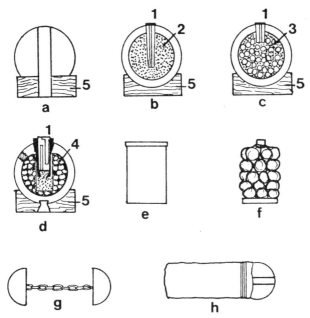

24. *Projectiles: a, round shot with sabot; b, common shell; c, shrapnel; d, Boxer's improved shrapnel and fuse; e, case shot; f, grape shot; g, chain shot; h, fixed ammunition: 1, fuse; 2, bursting powder; 3, bursting powder and balls; 4, balls and packing; 5, wood sabots.*

The beech-wood fuse had been replaced in the late 1830s by a number of pre-set fuses for various ranges. These fuses only required selecting for the required range and putting into the shell. In 1849 Captain Boxer introduced his time fuse (fig. 13f) which only needed opening with an auger in the hole corresponding to the time required and then inserting in the shell. In 1846 Quartermaster Freeburn invented the concussion fuse and in 1850 Commander Moorson introduced his percussion fuse.

The improvement in the construction of ships and the introduction of the ironclad meant that an armour-piercing shell was required and this was soon designed. Segmented shells, a cross between case shot and shrapnel, were introduced. Now that the windage was eliminated in shells a new fuse had to be designed, as it was the flash and windage that ignited the wood and Boxer fuses. For shells fitted with a gas check a wood fuse with detonator was introduced but was soon replaced by a time and concussion fuse. By 1912, after many years of trials and experiments, a time and percussion fuse was introduced that proved satisfactory.

The firing of the pieces had also improved. Instead of the loose powder poured on to the vent, small tubes of paper or goose feathers filled with the priming material had been introduced (fig. 13h). Another method, first used in the Royal Navy, was the use of gun locks (fig. 13b), but these were not used on all naval guns nor were they introduced into the land service until well after 1800 and then only for the larger garrison guns.

The detonating quill tube, introduced in about 1845, was designed to supersede both the gun lock and the linstock, as it required only a blow to set it off. In 1853 the detonating tube was replaced by a friction tube (fig. 13i) which was metal for the land service and quill for the Navy.

With the coming of the effective breech-loaders towards the end of the nineteenth century a big gun was loaded with the shell followed by bags of cordite and then the priming case which had in its base a primer which was fired by a percussion cap. This cap was activated by being hit by the firing pin in the breech block, which was worked by a trigger or lanyard. Once ignited the primer fired the bags of cordite which propelled the shell. Fixed ammunition of this type consisted of a case and shell in one rather like a giant rifle bullet (fig. 21d).

Until 1892 gunpowder had been the propellant and the bursting powder for shells, but in the six years that followed two new substances would be introduced that would revolutionise gunnery, cordite and lyddite. Cordite was used in propelling the shell. It was smokeless and required a fraction of the weight of the gunpowder to achieve the same results. Lyddite was a high explosive which burst shells. It was named after an experimental station at Lydd in Kent and was packed in special yellow-painted shells which had fuse holes in the top and were cast thick at the base in order to lessen the effect of the gun's discharge on the explosive content of the shell. Lyddite not only burst the shell but also acted as a powerful destructive force on the area where the shell landed.

7. MODEL MAKING

Before starting to make a scale model of a piece of artillery one must decide exactly what scale is to be used. This, of course, depends on what you are making the cannon for. If it is to be purely decorative then any scale will do, bearing in mind the finished size of the gun complete with its carriage. On the other hand, if the cannon is intended either for wargames or to fit into a diorama of model soldiers or sailors then a definite scale has to be chosen.

Most wargame models are 30 mm., but in some cases 20 mm., and so the cannon has to be made to this scale. A useful rule of thumb is to take the 20 or 30 mm. figure as representing an average-height man and then to scale down the cannon from there. If the gun is being made to complete a larger diorama, then 54 mm. is the scale. Once again the same rule of thumb can be applied as with the 30 mm. figures. The methods of making artillery are varied according to the size and to the facilities the maker has at his disposal. As the methods of making the smaller models differ from those used for the larger, decorative models, they will be dealt with separately.

In making a cannon of either 30 mm. or 54 mm. the most important part is the barrel. There are a number of different ways of making the barrel, depending on the facilities, or, and this is important for wargames, the number of cannon required. For brass ordnance the use of brass would be the most obvious but to turn barrels up from solid requires a lathe which is capable of slow running so that the contours of the barrel can be accurately cut. On the other hand brass tube of the correct size could be used for the barrel, with the rings made from brass wire and brazed or soldered on. The breech end can then be plugged with solid brass and shaped with files. The barrel can also be made from wood. This method is the least expensive and is possibly the easiest way. It also forms a good pattern for casting should a number of cannon be required for wargames. To make a metal barrel a mould has to be made. Using the original barrel made either in wood or brass make a plaster of Paris mould in two halves. Tie the mould tightly together and prepare the metal, which should be a mixture of lead and tin, as pure lead on its own will be too soft. When the metal has been melted it is poured into the mould and allowed to cool. One most important thing to remember is to pour in one go as any interruption will cause the metal already poured to set and the new pour will not fuse with it and you will have a barrel in two pieces.

The carriage should be made in wood, depending on the piece being made. With the smaller scales it is not possible to include a lot of the detail of the original, and so the maker must decide what detail he is including, bearing in mind the facilities available and his own capabilities. If the carriage is to be cast in metal from the original wooden model then great care must be taken to leave the carriages in pieces and to make separate moulds for each item.

The most difficult part of the field carriages will be the wheels (fig. 11). In an original model these should be made of wood and then, if desired, can be cast afterwards. Once one wheel has been made then the rest can be made in lead and you have a ready source of supply of wheels for any field carriage of your chosen scale. The making of wooden wheels is most difficult and time-consuming, and in the smaller scales takes an infinite amount of patience. A simple method of wheel-making is to cut out the wheel and its spokes in balsa wood from a sheet of suitable thickness. Once you have this pattern the wheels can be cast and then finished with a knife or file. The dishing can be done after the wheel is cast by applying a small amount of pressure in the centre of the wheel.

Rather than make models from scratch yourself, one can buy either ready-made models or kits. These come in various sizes in either metal, plastic or like the real thing.

The next type of model that can be made is the larger decorative model which is normally about 1/10 scale. For this the model maker requires a few more facilities than the maker of the smaller models. The most difficult part, as in the smaller guns, is the barrel. There are two alternative methods of making the barrel. It can be turned from solid and then bored, or it can be cast. The last method is probably the more satisfactory if you do not have the use of a suitably sized lathe. The casting method offers added advantages in that you can include more detail and decoration on the barrel. The original will have to be made in wood either by hand or with the help of a wood lathe and a power drill. Once the shape has been made the other detail can be put on. The crest of the reigning monarch and that of the Master General of Ordnance can be put in the correct places on the barrel.

To make these use either plaster, which is fragile, or cut them out of sheet brass with a metal fretsaw.

Once the model has been made it can either be cast in a mixture of lead and tin by making a plaster of Paris mould, or it can be cast in brass. The former method can be done by the maker himself and produces a suitable barrel which can be blackened with acid to give the appearance of a cast iron barrel. To have the barrel cast in brass the maker will have to know of a suitable fine-art brass caster. These are not too easy to find nowadays and the maker may have to send the barrel to Birmingham where there are a number of brass casters who would undertake this work.

The carriage, whether field or garrison, is made of wood. A good, fairly hard wood should be chosen and carefully cut to shape. It is useful to draw out the various pieces on paper before cutting so that you get the correct size, shape and angles. The metalwork for the carriages can be made from mild steel or in wood and then cast using plaster of Paris moulds. The latter method has the added advantage of having the parts available for future models. To make an iron garrison carriage or the iron trucks for a wooden garrison carriage the various parts should be made in wood before making a plaster mould, and then cast in metal. In some cases the model for casting can be made from modelling clay rather than wood.

The wheels of larger models have to be made as the originals were. Commencing with the hub or nave, which has to be turned to shape, the mortices are cut to take the spokes which should be made from square-sectioned wood and shaped before fitting into the slots in the hub. To get the dish of the wheel take a piece of wood about $\frac{1}{4}$ in. thick and cut a hole in the centre to take the hub. Place the hub in the hole with outer part inwards, and when the spokes are positioned the correct amount of dish will be given to the wheel.

When the spokes are firmly in place the felloes have to be fitted. These are best cut from a good-grained piece of wood as an entire wheel and then cut in sections. Draw out the exact size of the felloes with a compass and then cut them out of the wood. Perhaps the hardest part is getting the felloes to fit well and this depends on the accuracy of the spokes. Once the felloes are firmly in place the rim or tyre has to be made and fitted. A strip of mild steel of the correct width bent into the correct-shaped circle does very well. Make the circle of metal very accurately so that it is a tight fit on the felloes and keeps the wheel in shape.

Another form of artillery model is the one which actually fires. For this the casting and boring of the barrel has to be done with the highest accuracy. The barrel, by law, has also to be proved at either the London or Birmingham proof house and the owner has to be in possession of a firearms certificate for the cannon.

A more simple way of making model artillery for wargames is to make it in the flat, that is only two-dimensional. Select the type of gun that you are going to make and draw a plan of the elevation or copy from this book to the size you require. For the small 20 mm. scale, the outline will suffice. The next step is to cut this out in balsa wood of a suitable thickness, which will give you the model for the plaster of Paris mould. Once the mould has been made the guns can be cast. A point to remember when making the model for the mould is that it will need a base on which to stand. When the model has been cast it can be cleaned up and painted.

Another quick method of making guns is to make them in the flat from either wood or cardboard. Once a template has been made the wood or card is cut to shape, fitted with a suitable base, and painted. If you have recourse to any suitable photographs of guns these can be stuck on card and cut out.

If painting on metal the paints used for model soldiers should be used. If the gun is liable to be handled a lot then it should be sprayed with lacquer. For any wooden parts of carriages there is a stain and wax combined that, having been painted on, requires only a rub with a duster to bring it to a brilliant shine.

GLOSSARY

Artillery: machines that hurl projectiles by explosive or other means.

Ballista: ancient artillery machine for throwing stones.

Barrel: the hollow tube of a firearm.

Bombard: the name of an early artillery piece.

Block trail: the trail of a carriage which is made out of one piece of wood.

Bore: the diameter or the length of the inside of the barrel.

Breech: the rear end of the barrel, which in the case of breech-loaders has an opening for loading.

Cannon: from *canae* meaning a tube. Used to describe heavy pieces, not hand guns.

Carriage: the frame that supports the barrel of a cannon.

Felloe or **felly:** a piece of wood that formed part of the wooden tyre of a wheel.

Gun: a piece of artillery that discharges a projectile by means of gunpowder.

Howitzer: a short gun firing shells at high angles.

Igniter: device or apparatus for igniting the powder.

Limbering: the term used in describing the action of attaching the gun to the limber.

Mortar: a short heavy piece for firing shells at high angles.

Muzzle: the front opening of the barrel.

Ordnance: a term to describe all manner of artillery.

Piece: a term used to describe artillery.

Q.F. or **Q gun:** a quick firer gun which has shell and charge joined together for quicker loading.

Quoin: a wooden wedge used in the elevating of barrels.

Recoil: the driving backwards of a gun after discharge.

Sabot: a wooden shoe strapped or riveted to a round projectile to prevent it turning in the bore.

Train: the collective term for guns, wagons and other impedimenta.

Windage: the difference between the diameter of the bore and that of the projectile which allows gas to escape.

BIBLIOGRAPHY

Baker, Harry. 'The Crisis in Naval Ordnance', *Maritime Monographs and Reports* number 56 (1983). National Maritime Museum.

Douglas, Sir Howard. *Naval Gunnery*. John Murray, 1819. Fourth edition of 1855 reprinted by Conway Maritime Press, 1982.

Garbett, Captain H. *Naval Gunnery*. London, 1897. Reprinted by S. R. Publishers, East Adesley, Wakefield, West Yorkshire, 1971.

Hogg, I. V., and Batchelor, John. *Naval Gun*. Blandford, 1978.

Hogg, I. V., and Thurston, L. F. *British Artillery Weapons and Ammunition, 1914-1918.* Ian Allan, 1972.

Hughes, Major General B. P. *British Smooth-Bore Artillery*. Arms and Armour Press, 1969.

Hughes, Major General B. P. *Firepower*. Arms and Armour Press, 1974. (Weapon effectiveness 1630-1850.)

Wilson, A. W. *The Story of the Gun*. Royal Artillery Institution, Woolwich. Latest edition 1985.

INDEX